MIX
Papier aus verantwortungsvollen Quellen
Paper from responsible sources
FSC® C105338

Israel Yigezu

The Challenges and Prospects of Potable Water Mangement at Edja Woreda- Ethiopia

Anchor Academic
Publishing

Yigezu, Israel: The Challenges and Prospects of Potable Water Mangement at Edja Woreda- Ethiopia. Hamburg, Anchor Academic Publishing 2013

Buch-ISBN: 978-3-95489-110-8
PDF-eBook-ISBN: 978-3-95489-610-3
Druck/Herstellung: Anchor Academic Publishing, Hamburg, 2013

Bibliografische Information der Deutschen Nationalbibliothek:
Die Deutsche Nationalbibliothek verzeichnet diese Publikation in der Deutschen Nationalbibliografie; detaillierte bibliografische Daten sind im Internet über http://dnb.d-nb.de abrufbar.

Bibliographical Information of the German National Library:
The German National Library lists this publication in the German National Bibliography. Detailed bibliographic data can be found at: http://dnb.d-nb.de

Hermannstal 119k, 22119 Hamburg
http://www.diplomica-verlag.de, Hamburg 2013
Printed in Germany

Table of Contents

Acronym

ADB African Development Bank

AIDS Acquired Human Deficiency Syndrome

CBOs Community Based Organizations

CSOs Civil Society Organizations

DME Design Monitoring and Evaluation

EUWI European Union Water Initiatives

FGD Focus Group Discussion

GDP Gross Domestic Product

GWI Global Water Institute

HIV Human Immune Virus

IFAD International Fund for Agricultural Development

MDG Millennium Development Goals

MDM Major Donor Ministry

MoFED Ministry of Finance and Economic Development

NGOs Non Governmental Organizations

SIWI Stockholm International Water Institute

SNNPRS South Nations and Nationalities and People's Regional State

SNV Netherlands Development Organization

SPSS Statistical Package for Social Science

UN	United Nations
UNDP	United Nations Development Program
UNECA	United Nations Economic Comission for Africa
UNFPA	United Nations Population Fund
UNICEF	United Nations Children's Fund
UNU	United Nations University
USAID	United States Agency for International Development
WaSH	Water, Sanitation and Hygiene
WB	World Bank
WSP	Water Supply Program
WGF	Water Governance Facility
WHO	World Health Organization
WMEO	Water, Mineral and Energy Office
WVE	World Vision Ethiopia

List of tables

Page

List of figures

Page

Abstract

This study was motivated by the beliefs that analyze ***The Challenges and Prospects of Potable Water Management at Edja Woreda.*** *Ethiopia in general and Edja Woreda in particular has lowest access to potable water services. The objectives of the study were; to describe the role of potable water actors and the networking status among them, the status of potable water management in terms of its supply capacity and distributive mechanisms, the institutional, economical and social mechanisms in place to contribute its sustainability and to generate suggested solution for potable water management problems. Mixed social science research was used for this study. The necessary data was collected through survey questionnaires; interview and FGD form potable water users, water actors' officials and selected community members. Descriptive statistics like: graphing, frequency and percentage distribution was used for data analysis using SPSS version 13 software.*

The study also reveals that the major potable water actors in the area are; WMEO that coordinates and manages over all water activities, WaSH committees that manage and monitor water scheme, NGOs and multilateral agencies that finance water activities. Water actors are both responsible and accountable to the users but needs them more improvement. Nevertheless, the networking status among them is good. The water management challenges in the area are; poor supply capacity and distributive mechanisms with problems like water interruption, limited supply capacity, and distances of water points. The study result also shows that the institutional, economical, social mechanisms are in place but lacks effectiveness to ensure sustainability. These results imply that there are some prospects of potable water management that needs more improvement.

Key Words; Potable water actors, Responsibility, Accountability, Institutional, Economical and Social Mechanisms

Acknowledgments

First and foremost, I would like to express my sincere gratitude to my advisor Dr. Paul Mukasa (PhD) for his continuous and tireless professional guidance throughout the proposal development, questionnaire design, data collection and write up of the research project. He earns especial appreciation for the input he added on my work in the entire stage of thesis writing.

Besides, I owe my deepest thanks to Dr. Menhaj Alem for the invaluable comments and suggestions he has made for the study. He has played unreserved role in the proposal development stage of the study.

Moreover, I am sincerely grateful for the cooperation made to me by World Vision Ethiopia Guraghe Area Development Program staffs, Edja Woreda water, and Mineral and Energy Office officials for the invaluable support they made to me during the data collection stages of the study. It would have been difficult to complete data collection without their cooperation.

Furthermore, I would like to thank World Vision Ethiopia WaSH division for sponsoring the study; without which the completion of this study would not have been possible in such a successful way. Besides, I would like to thank the division for the proper arrangement of work time for the study.

Finally but most significantly, I would like to express my deepest thanks to my family and friends for their encouragement during the entire study time. I especially indebted to DME team of my colleagues for the empowering role they played for me during entire study time.

CHAPTER ONE

INTRODUCTION

This chapter introduces the back ground of the study, the statements of the problem, objective of the study, research questions, and scope of the study, limitation of the study, significances of the study and organization of the paper.

1.1 Background of the study

About 97 percent of the earth's water exists in ocean form which is salty and not used for drinking purpose unless treated costly. Only three percent of the earth's water is fresh water. From the three percent, only 0.3 percent exists in accessible (river water, ponds and lakes) form since 99.7 of it exist in inaccessible form; such as ground water, glaciers and icecaps (Izquierdo et al., 2004).

On the other hand, "about 1.6 billion of the world populations live in areas of economic water scarcity where the unavailability of human, institutional and financial capital limits access to water; even though adequate unutilized water is in fact available to meet human needs" (Mwebaza, 2010, p. 5).

Furthermore, about 5,000 children die from preventable water related diseases every day on our planet (UNDP et al., n.d). Similarly, Potable water coverage in Sub-Saharan Africa remains below 60 percent of the population (UNICEF, n.d). Thus, more than 250,000 children die in Ethiopia each year because of poor sanitation and hygiene (UNDP and SNV, 2009).

Ethiopia is one of the nations blessed by water resource on the world having 12 major river basins and 12 large lakes. The annual surface water runoff is estimated to be 122 billion m^3. Besides, the country has an estimated 2.6 billion m^3 of usable ground water. On the other hand, Ethiopia has facing those challenges like the dynamics of population growth, low potable water access, low productivity, structural bottlenecks, dependence on unreliable rainfall and being land-locked which are all working against the country's economic wellbeing. The addition of an estimated 2 million persons per year puts tremendous pressure on the country's resource base, the economy and the ability to deliver social services (MoFED, 2005).

Thus, Ethiopia has the lowest access to potable water supply in Africa despite its abundant water resource (USAID, n.d). The national level access to safe water in Ethiopia is 68.5 percent of the total population with 91.5 percent and 65.8 percent for urban and rural potable water access respectively. The average safe water coverage in SNNPRS is 62% which is 90.9% and 58.7% in urban and rural areas respectively (ProAct, 2011).

In the Edja *Woreda* of Guraghe Zone, there are 32 on spot springs; two hands dug wells and one water supply system by gravity called Fessa water supply which makes up the water supply of the *Woreda* 43% of the total population.

1.2 Statements of the problem

Potable water is the water delivered to the consumer that can be safely used for drinking, cooking and washing (De Zuene, 1997). In the Edja *Woreda* of Guraghe Zone, the average distance travelled to collect water is 46 minutes. On the other hand, the average water consumption is 7.2 liter/day per person for only 48 percent of the population while the remaining 52 percent consumes less than 7.2 liter/day/person. These figures are far below the average daily water consumptions of 45 litter/person per day set by WHO and the 20 litter/person per day of Millennium Development Goal of universal access program (ITAB Consult, 2011).

Water born diseases are common in the *Woreda* especially those affecting children. Even in some area of the *Woreda* where water schemes are available, the utilization of potable water schemes creates wastage and unsustainable. One of the major potable water related problems in the *Woreda* is maintenance of water schemes. The need for maintenance report lags behind expected in case of scheme damage. Water fee collection in the *Woreda* is also ineffective. There is no clear responsible actor in maintaining water schemes. Hence existing WaSH actors such as; institutions, legal frameworks, and policies are in question to ensure the sustainability of expanding access to drinking water while preserving environment. There is poor integration among actors to solve water related problems (WVE, 2011).

The financial sustainability of the drinking water sector in the *Woreda* remains an issue for all stakeholders: the providers, the users, government agencies and donors. There is no clear guidance for water activities at *Woreda* level. Thus, there are no clearly specified roles and responsibilities. Even if access to water and water education is provided to the residents of the study area, practical usage of water lag far behind the universal millennium development target in the Edja *Woreda* (WVE, 2006). The global drinking water crisis is mainly rooted in poverty, power and inequality, not in physical availability. It is, first and foremost, a crisis of governance (Harris et al., 2011). Based on these facts, the study tries its best to explore whether management gap that halted the expected progress on the potable water services delivery in the study area.

1.3 Objective of the study

The general objective of the study is to identify the challenges and prospects of potable water management in terms its supply capacity and distributive mechanism in the Edja *Woreda* with the following specific objectives;

1. To analyze the role of potable water management actors (Woreda Water, Mineral and Energy Office, NGOs, WaSH Committee and the beneficiaries) in the study area in terms of responsiveness and accountability.
2. To analyze the networking status between major potable water actors (Woreda Water, Mineral and Energy Office, NGOs, WaSH Committee and the beneficiaries) in Edja *Woreda*.
3. To analyze the status of potable water management in Edja *Woreda* in terms of supply capacity and its distributive mechanism.
4. To assess the challenges and prospects of institutional, economical and social mechanisms in place to contribute the sustainability of potable water supply in Edja *Woreda.*
5. To generate suggested solutions for the enhancement of better potable water management in Edja *Woreda.*

1.4 Research questions

The study endeavors to explore the prospects and challenges of potable water management in Edja *Woreda* by answering the following questions;

1. What is the role of potable water management actors (Woreda Water, Mineral and Energy Office, NGOs, WaSH Committee and the beneficiaries) in the study area in terms of responsiveness and accountability?

2. What is the networking status between major potable water management actors (Woreda Water, Mineral and Energy Office, NGOs, WaSH Committee and the beneficiaries) in Edja *Woreda*?

3. What are the status of potable water management in Edja *Woreda* in terms of water supply capacity and its distributive mechanism?

4. What are the challenges and prospects of institutional, economical and social mechanisms in place to contribute the sustainability of potable water supply in Edja *Woreda*?

5. What are the possible solutions for the enhancement of better potable water management in Edja *Woreda?*

1.5 Scope of the study

Edja *Woreda* of Guraghe Zone has 28 *Kebele* administrations which cover an area of 767 km^2 in which more than 84 thousand people dwell. Due to the cost and time limit constraints to conduct the study in the whole *Kebeles* of the *Woreda* with proper sample size, this research is limited to the study of potable water management in selected six sample *Kebeles* in the *Woreda*. Thus, the study result does not reflect the entire Guraghe Zone potable water management situation. On the other hand, the word management is a broad term by itself that the study only reflects facts about the challenges and prospects of potable water management in the study area. Besides, Potable water is the water delivered to the consumer that can be safely used for drinking, cooking and washing that the study is limited to this definition of potable water. Thus, the study is limited to the challenge and prospects of potable water (water scheme and water supply) management in the study area in terms of; water supply capacity and its distributive mechanism, role of potable water management actors and the status of networking among them.

Furthermore, this study is also limited to assess the status of institutional mechanism (water regulatory bodies, management arrangement, planning procedures), economical mechanism (water pricing, charging and water related penalties) and social mechanisms (measures to increase awareness of water issues and mobilizes users to participate in planning, management and financing of water resource development) in place to ensure sustainability.

1.6 Limitation of the study

Most of the residents of Edja *Woreda* are rural dwellers and live in scattered areas. Hence, it is difficult for the researcher to deal with all respondents and they are to be contacted by enumerators. Since the respondents are not well aware about the research, they were unable to give appropriate information to data collector unless they are convinced very well. Since everyone have his/her own daily business, getting the right respondent with sufficient time may create difficulty in the process of data collection. To avoid these limitations and other unforeseen obstacles, training was given to data enumerators beforehand which help them to extract sufficient information for the study.

1.7 Significance of the study

Most of the structural bottlenecks of developing countries are interpreted with financial capacity and physical constraints. However, the issue of management matter most in having quality water services. Thus, the study result plays great role in finding alternative solutions for the improvement of water management in the study area. The study can also benefit the community of the study area in identifying those problems that hinder the efficiency and sustainability of water program. The finding can also be used as additional reference for government officials and NGOs that are working on water program in providing them additional knowledge about potable water management.

1.8 Organization of the Paper

The research project is organized in such a way that the first part of the paper highlights the background of the study, the statement of the problem, objectives of the study, scope and delimitation of the study. Chapter two addresses some empirical evidences about the subject under study. Chapter three deals with the study methodology while chapter four broadly explains about the finding of the study. Finally, chapter five summarizes the finding and draws some helpful recommendations on potable water management.

CHAPTER TWO

LITRATURE REVIEW

In this chapter the concept of water potable water management term and the basic water supply were defined. Besides, secondary documents about the potable water management was reviewed and interpreted with the finding of the research in the analysis part of the study. Finally, the contribution of potable water supply for the MDGs was summarized and conclusion was made for the entire literature review.

2.1 Conceptualizing water management

The basic water supply is defined as the provision of effective water use as well as a minimum quantity of 25 liters of potable water per person per day (or 6 000 liters per household per month) within 200 meters of a household, which is not interrupted for more than seven days in any year; and with a minimum flow of 10 liters per minute in the case of communal water points. Potable water is defined as drinking water that does not impose a health risk (Khambule, 2002).

Potable Water management is the activity of planning, developing, distributing and managing the optimum use of drinking water resources. In an ideal world, water management has regarded to all the competing demands for water and seeks to allocate water on an equitable basis to satisfy all uses and demands. Potable water supply services means the concept from a water resource, transportation, treatment, storage and distribution of potable water, water intended to be converted to potable water and water for industrial or other use, where such water is provided by or on behalf of a water services authority, to consumers or other water services providers (GWP et al., 2008).

2.1.1 Modalities of water service delivery

There are five basic modalities of potable water management service delivery; namely: public provision (when the central or local government directly provide water services), private provision (when private company provide water services and charge clients), community based provision (when local community provide their own water services), public private partnership (when both public and private entities jointly provide water services) and multi stakeholder provision implies when a number of actors recognize their common water management problems and realize the value of their collective to solve those problems. Thus, in case of multi stakeholder water services provision is made through multi stakeholders (Miranda et al., 2011).

2.1.2 The tragedy of the commons

Human beings are constantly making use of goods, natural resources, and spaces for consumption and waste disposal. The use of these goods, resources and space are also available to other users in many cases. However, the rational individual concludes that the benefits of the common can be enjoyed, without causing any more than the slightest damage to it. This rational individual conclusion and action is viewed by Garrett Hardin as the "the tragedy of the commons" (Christensen, 2005).

According to Hardin (1968), the tragedy comes as the usage of each common go beyond the optimal usage level. Hardin also applied his view to the situation created by pollution and quoted as "The rational man finds that his share of the cost of the wastes he discharges into the commons is less than the cost of purifying his wastes before releasing them. Since this is true for everyone, we are locked into a system of "fouling our own nest," so long as we behave only as independent, rational, free-enterprisers" (Hardin, 1968, p. 1245).

A common-pool resource like lake or ocean, irrigation system, forest, internet, the atmosphere, are natural or man-made resources from which it is difficult to limit or exclude once the resource is available for users. Accordingly, when a person consumes units of these resources, he/she remove those units available to others. This fact is applicable to both renewable natural common pool resources and man-made common pool resources. when a unit of common pool resources have high value with the absence of institutional restrict to limit the way resource units are appropriated, individuals get strong incentives to appropriate more and more resource units leading to congestion, overuse, and even the destruction of the resource itself. Thus, the free-rider problem is a potential threat to efforts that reduce appropriation and improve the long-term outcomes achieved from the use of the common pool resources (Ostrom, 1999).

2.1.3 Good governance

In the 1990s, the term governance has attracted multilateral agencies like World Bank from the concern of the sustainability of projects financed by those agencies. The term governance particularly good governance which is characterized by different elements that are pointed out by different institutions based up on their own perception highly correlated with the notion of sustainable development (IFAD, 1999).

The former UN secretary general, Kofi Annan, has also bolded the importance of Good governance in achieving the eight Millennium Development Goals (MDGs) by quoting as "Good governance is perhaps the single most important factor in eradicating poverty and promoting development" (Abdellatif, 2003).

Good governance comprises the existences of effective mechanisms, processes and institutions that enable citizens and groups to articulate their interest, exercise their legal right, meet their obligations and mediate their differences. Good governance is characterized by those elements such as: participation, rule of low, transparency, responsiveness, and consensus orientation, equity, efficiency, effectiveness, and accountability, coherent and integrative (Rogers and W. Hall, 2003).

2.2 Empirical literature review

The global burden of the inadequate access of safe water falls primarily on the poorest of the poor. There are four universal barriers that hinder the progress of safe water services supply. These are: 1) inadequate investment in potable water infrastructures, 2) Lack of political will 3) difficulty in acceptance of new technological approaches and 4) failure to conduct evaluations of water interventions. Equity and financial sustainability have good link with one another. Water services needs user fees and inputs from beneficiaries for its sustainability in the form of water fees or contribution of time and money to initial project establishment. It helps in such a way that services are adequately valued, maintenance is provided, overuse of scarce resources is avoided, and limited external resources can be prolonged as much as practical (L. Moe, 2006).

While the world is working toward the achievement of MDG target of safe water supply, there are still major gaps in many regions and countries especially in Sub-Saharan Africa. This is mainly due to poverty, inequality, and lack of physical availability which is all the subsets of mismanagement. Thus, good potable water management is a key pillar of effective and efficient water service provision (UNDP et al., n.d).

The current deficiancy of potable water access in many develping countries is global fact. Developing countries are also adimited that they cannot solve their deep rooted potable water problems working alone and their governance system must permit all stakeholders to actively engage and solve the growing potable water problems. The current potable water crisis requires nations to put their governance systems in order; since sustainability and development are inseparable under present global conditions. More effective management systems need to be created to overcome government failure, market failure and system failure or a combination of these (Rogers and W. Hall, 2003).

The effort to decrease the numbers of people that lack adequate access to safe water in Africa in general and in Sub-Saharan Africa in particular has proved to be significant challenges. Hence, the region is lagging behind the universal target of MDGs that aim to half the proportion of people that has no safe water access by the year 2015 (ADP et al., 2006).

African counries are remain charcterized by poor quality and lack of availabilities of basic infrustructures like; electricity, water and sanitation, and roads are well below other regions. Water service provision in Africa is known by its inefficiency and the lack of pricing mechanisms to determine consumer demand and to reflect service costs (GWP et al., 2008).

Africa is unlikely to reach the MDG target of achieving universal drinking potable water services by the 2015. Serious reforms in institutions, legal frameworks, and policies are needed in order to ensure the sustainability of expanding access to drinking potable water while preserving environment to achieve MDG target in potable water supply (Kauffmann, 2007).

Decentralization should be made in many developing nations to solve potable water crises along with necessary financial resources and human capacity development at local level through clear demarcation of roles and responsibilities. All stakeholders, from communities to regional partners, should participate in the management of water resources to improve efficiency and avoid conflict. Community level involvement is also important to overcome local environment and development confilicts, property rights, equity and litracy issues while the participation of local government, non traditional palyers, and water oriented civil societies are also vital through strengthening local water committee and creating efficient and effective public water resource management (WSP, n.d).

Ethiopia's potable water statistics have been considered among the lowest in the world in terms of access, distance and quality. On the other hand, Ethiopia has one of the countries that have made significant progress toward the achievement of MDGs target of potable water through radical program reform and decentralization. In the year 2005, the government of Ethiopia has launched a water and sanitation plan that aimed to achieve 100% water and sanitation access throughout the country by the year 2012 (WB et al., 2009).

The institutional frame work of Potable water in Ethiopia includes: Ministry of Water Resources, Ministry of Health, the Regional Water Bureaus, Regional Health Bureaus, *Woreda* Water Desks, and the Ethiopian Social Rehabilitation and Development Fund (ESRDF), private sector in the form of contractors and consultants, equipment and spare parts suppliers, and water vendors, artisans, communities with support from NGOs and donors (ADB, 2005).

Safe water supply access in Ethiopia is supported by dozens of donors and hundreds of NGOs each working by their own criteria and procedures. To tackle this donor fragmentation, the government has established a multi stakeholder forum that includes donors, NGOs and Civil Society Organizations (CSO). The government has also developing computerized monitoring and evaluation program for water supply, sanitation and hygiene program consultancy services in order to solve data handling problems (UNDP and SNV, 2009).

Edja *Woreda* has great challenges of maintenance systems in place that result in inadequate services (43% of the total population). Most of health institutions and schools in the area have no water supply system and hand washing facilities. Besides, the community of the area has low level of understanding and awareness on knowledge, attitude and practices of potable water (ITAB Consult, 2011).

Potable water management in the study area is undertaken by Edja *Woreda* Water, Mineral and Energy Offices. Water system in the study area is managed by water, sanitation and hygiene committee with the close consultation of *wareda* Water, Mineral and Energy Office. The community participates in potable water programs/projects in site selection, providing labor and local materials. Other WaSH actors of the area includes: Kale Hiwot Church, UNICEF, WB, and WVE (Ibid).

2.2.1 The link between safe water supply and the Millennium Development Goals (MDGs)

MDGs are internationally agreed targeted goals to tackle some of the major development challenges within the time range of 15 years. It comprises eight major goals namely: 1) Eradicate extreme poverty and hunger, 2) Achieve universal primary education, 3) Promote gender equality and empower women, 4) Reduce child mortality, 5) Improve maternal health, 6) Combat

HIV/AIDS, malaria, and other diseases, 7) Ensure environmental sustainability, 8) Develop a global partnership for development. Each of these goals has different set targets that directly contribute for the achievement of that goal (Rao, 2006).

Potable water supply is an integral aspect of achieving almost all the Millennium development Goals (MDGs). Access to safe water has strong correlation with the MDGs of number: 1, 3, 4, 5 and 6 (UNU, n.d). The use of unsafe increases child mortality; endanger maternal health; prevents children from going to school due to water born diseases; and worsens poverty. Clean water is a proven cost-effective investment that was reduces health budget spending and increases economic productivity and education levels (UNDP et al., n.d).

The relationship between improved water and the mellienium development goals is briefly summarized in table 1. below.

Table 1. The relationship between Potable water and MDGs

S/N	MDG theme	Contribution of improved Potable water services to the MDG
1	Eradicate extreme poverty and hunger	▪ Household livelihood security rests on the health of its members and access to potable water. Access to safe water also improves nutrient uptake. ▪ Access to safe and affordable water fees helps the fetchers of water to engage in productive activities.
2	Achieve universal primary education	▪ Improved potable water supply services relieve children, especially girls, from water fetching duties, allowing them to attend school (easy access to water was give girls and boys more time to attend school) and reduce water and sanitation related disease. ▪ Pure water and sanitation services results in different water related diseases which increase children absenteeism from school.
3	Promote gender equality and empower women	▪ Potable water access helps empower women directly and indirectly. Reduced time, health, and care-giving burdens from improved water services give women more time for productive endeavors, adult education, empowerment activities, and leisure. ▪ Without safe and close-by water facilities, girls risk sexual harassment in toilets and river or on their way to them.
4	Reduce child mortality	▪ Safe drinking water sources and greater quantities of domestic water for washing reduce infant and child morbidity which are the single biggest killer of children in the developing world.
5	Improve maternal health	▪ Accesses to pure water services reduce labor burdens and health problems reducing maternal mortality risks.
6	Combat HIV/AIDS, malaria, and other diseases	▪ Access to water is the foundation of health and therefore prerequisite for ameliorating major diseases which can lead AIDS patients to mortality.
7	Ensure environmental sustainability	▪ Since water is the natural recipient of most pollution, adequate treatment and disposal of excreta and wastewater contribute to less pressure on freshwater resources and aquatic animals. ▪ Inadequate access to safe water is one of the five defining characteristics of a slum.
8	Develop a global partnership for development	▪ Negligible or very weak relation

Source: (Rao, 2006)

2.3 Summary of litrature review

Potable Water management is the activity of planning, developing, distributing and managing the optimum use of drinking water resources. There are five different modalities of water services namely: public provision, private provision, community based provision, public private partnership and multi stakeholder provision.

The empirical study reveals that the insufficient supply and lack of physical availability of water on our planet is the subset of water mismanagement. Thus, the governance system of a country must permit all stakeholders to actively engage and solve the growing potable water problems through the establishment of effective water management. Community level involvement in water system development is also important to overcome local environment and development confilicts, property rights, equity and litracy issues while the participation of local government, non traditional palyers, and water oriented civil societies are also vital through strengthening local water committee and creating efficient and effective public water resource management.

Water user fees and inputs from beneficiaries during the initial project establishment helps: to value services adequately, to provide maintenance, to avoid overuse of scarce resources, and to prolonged water services as much as practical.

According to the empirical study, there is the lack of strong and participatory water management system in the world which is also one of the common potable water service problems in the study area. Efficient and effective water management system should be in place to ensure sustainable potable water supply.

Potable water supply in Africa in general, and in Sub-Saharan Africa in particular has been considered among the lowest in the world in terms of access, distance and quality. Similarly, Edja *Woreda* of Guraghe Zone has the lowest water supply coverage that only 43% of the population has access to safe water supply. Most of the schools and health posts in the *Woreda* have no water supply system.

CHAPTER THREE

METHODOLOGY

This chapter describes the methodology used in the fourth chapter of the research paper. Thus, the theoretical framework, the conceptual framework, operational definition of variables, description of the study area, research design, methods of data collection, instrumentation, sampling methods, methods of data analysis, ethical consideration were discussed in this chapter.

3.1 Theoretical framework

There are a number of theories that are applicable to water management related studies. Nevertheless, only the network management theory was used in this study.

Network can be defined as bounded sets of actors, organizations, or individuals that are connected by specific relationships. It refers to the study of structural forms of patterning of the ties that link these units. Governance networks are a particular form of governance which includes a relatively stable horizontal articulation of interdependent, but operationally autonomous actors that interact through negotiations which involve bargaining, deliberation and intense power struggles. The governance network takes place within the institutionalized frame work of articulated rules, norms, knowledge and social imaginaries. This is a self regulating network that takes place within limits set by external agencies that contribute the production of public purpose (Sørensen and Torfing, 2005).

Network theories of governance include approaches that investigate patterns of interest intermediation and public private cooperation in the making and implementation of public policies. The common concern is on how these actors and agencies come to form networks, what holds them together, what determines their choices and how they influence political and social decisions (Toikka, 2011).

3.2 Conceptual framework

Net work theory leads to the conceptualization of the study in that there must be networking relationships to achieve the desired results in potable water management. Water governance refers an emerging governing practice which seeks to build greater capacity for collective action through new integration between diverse social actors. Thus, it focuses on multi agencies' actions such as partnerships and networks developed for creating synergy among different social actors in the pursuit of public policy goals. In the society centric approach of governance, alternative modes are less reliant on top down policy instrument but need to find out more accountable and democratic means of social organization where responsibility and accountability for intervention is shared equally between public, community and private actors. In the governance process, there are three major actors mainly, the state, the private and civil society (Allen et al., 2006).

Resource management matters a great deal for the economic, social and environmental outcomes. This means that when there is good governance, policies, programs and projects can yield the targeted goals or outcomes. On the other hand, poor governance resulted in the political and social risk, institutional failure and rigidity and deterioration in the capacity to cope with shared problems which means failing to achieving sustainable development outcomes. Poor management is thus, an obstacle to development while good governance the most important factor to poverty reduction (Rogers and W.Hall, 2003).

Therefore, the actors of potable water management should be responsive, accountable and have to have good networking among them to achieve good potable water management. Besides, potable water management instruments such as: institutional, economic and social mechanisms should be in place in order to ensure the coordination, planning procedures, efficient and effective allocation of safe water, create awareness of water issues and mobilize users to participate in planning, management and financing of potable water resource development. Adequate access of water and sanitation in turn contributes to the achievements many MDGs that lead to development. The following figure summarizes the overall relationship between the different variables of resource management that affect good potable water management both negatively and positively.

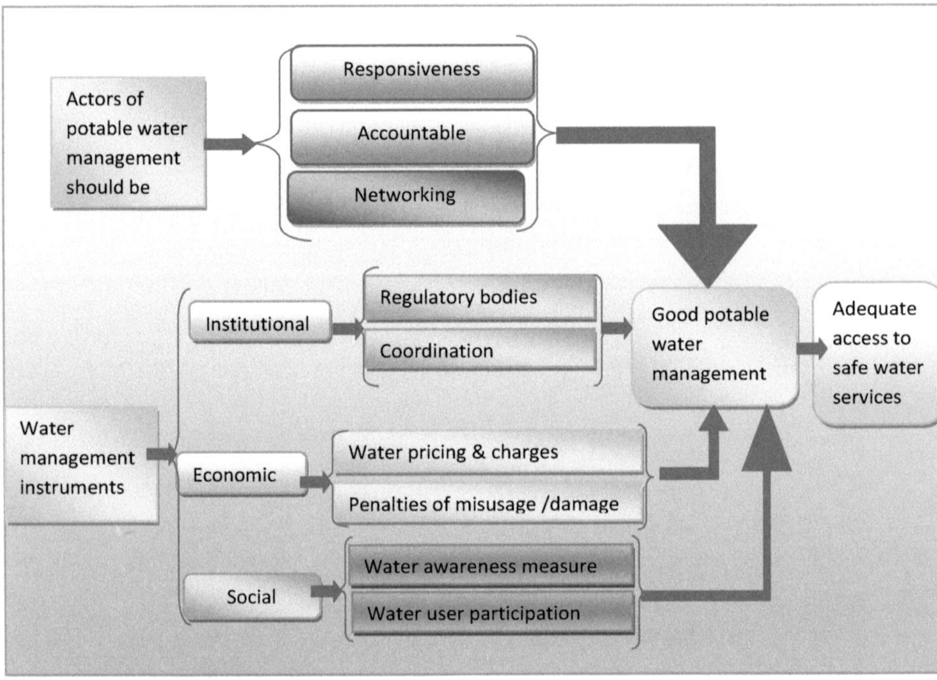

Figure 1. Conceptual framework

3.3 Operational definition of variables

The variables for this study were adequate access to potable water services, household level water consumption and quantity of water supplied which was measured as quality and timely services, household water consumption per day in liter and potable water supply capacity in liter respectively. The responsiveness of actors, accountability of actors and networking among potable water actors, existence of water regulatory bodies, coordination and partnership among actors, water pricing and charging mechanisms, penalties, water awareness measures, and water user participation were also the variables of the study. Responsiveness refers the institution and process should give appropriate and timely services to all stakeholders while accountability refers decision makers should be accountable to the public and as well as to the institutional stakeholders. A regulatory body refers those who enforce water issues, coordination and partnership which indicate the degree of coordination and partnering between different water actors.

Water pricing and charge means user fees and charges imposed on users in case of water misusing and scheme damage. Water awareness measure means measures used to increase awareness of potable water issues while user participation indicates participation of users in planning, management and financing of potable water resource development (Harris et al., 2011). Responsiveness was measured as the time (days) that takes potable water actor to respond to the demand of the beneficieries while accountability is measured as the degree to which actors were responsible for the action or decision they undertake. On the other hand, networking among potable water actors was measured as the number of potable water activities that implimented with out deuplication of efforts and through mobilization of various resources from different actors.

This was where network management theory was applied to evaluate the integration among actors to achieve drinking water goals, to what extent they invite water users and stakeholders in decision making, how effective, efficient, responsive and accountable they were to achieve the common potable water services.

3.4 Description of the study area

Guraghe Zone is one of the Southern Nations and Nationalities People's Regional State (SNNPRS) which is located in the South Western part of Ethiopia and in the Northern tip of the SNNPRS which is bordered by Silte and Alaba Zones and Yem special *Woreda* in its Southern part and Oromia regional state in its Northern, Western and Eastern part. The Zone encompasses an area of 7,215 square km (WVE, 2006).

Edja *Woreda* is one of the 13 *Woreda* of Guraghe Zone that is located in the Southern Western part of Ethiopia on the distance of 195KM from the capital city Addis Ababa which covers 767 km^2 with the total population of 84,882 out of which 40,247 male and 44,635 female indicating a percentage of 47.41% and 52.59% male and female respectively. The annual population growth rate of the Woreda is estimated to be 3% (Ibid, 2008). From the total area of the *Woreda,* 40.7% comprise cultivated land while the pasture and forest land covers 12.1 and 8.6 percent respectively. (Nigatu, 2011) The *Woreda* has 28 *Kebele* Administrations (KAs). It has three different agro-ecological Zones namely; high land or *Dega* which is 53% of the total area, mid highland or

Weina Dega (37%) and low land or *Kolla* (10%) with annual mean temperature of 19^0C and its average rainfall is 1000-1250mm per annum (Melissa and Janson, 2008).

Agriculture is the dominant sector of the *Woreda's* economy and hence more than 80% of the livelihoods of the population depend up on mixed farming. The major crops grown in the area are; *enset*, barely, potato, maize, field pea, *teff*, chat and coffee. *Enset* is the major food of the *Woreda*. There are three different soil types in the *Woreda*, 12% black, 50% red, and 37% brown soil. As it is common in many *Woreda* of the country, deforestation is one of the major environmental problems of the Edja *Woreda*. As a result, soil fertility and hence ground water is being depleted gradually (WVE, 2006).

3.5 Research design

Mixed social science research which use primarily and dominantly qualitative research approach was used in this study. Questionnaire was designed and distributed to the beneficiaries of the potable water services in the area, Water, sanitation and hygiene committee and government officials from which potable water management related raw data was generated. Secondary documents from the potable water actors in the study area were also reviewed to get supplementary data. The target population for this study was the beneficiary of potable water services, WaSH committee as well as the experts from government officials (health extension workers, *Woreda* Water, Mineral and Energy Office officials) in the Edja *Woreda*.

After data collection was completed, all the collected data was checked and verified for its validity. The questionnaires were coded to make data ready to feed in the Statistical Package for Social Science (SPSS) software. Accordingly, a computer based data dictionary and data entry system was developed for survey questionnaires using SPSS. A random visual comparison between the original questionnaire forms and the computer database was undertaken after the completion of data entrance. The SPSS data base program version 13 was used to process raw data in order to generate the needed information from which interpretation was made to summarize research findings using descriptive statistics.

3.6 Methods of data collection

The data was collected from 6 selected *Kebeles* in the *Woreda* namely: Sheramo, Shebreden, Desene, Agena 01, Weyra Debane and Wasamar. Except Agena 01 which is small town, all the population of sample *Kebeles* are rural inhabitants and exist in scattered form. From each of these *Kebeles*, different numbers of potable water beneficiaries was selected by proportionate stratified random sampling since the *Kebeles* have different population size. Thus, a total of 265 water users were contacted to collect raw data. Enumerators were trained to ensure the quality of data collected by them. The necessary data was collected from these selected samples through questionnaire. Besides, data was also collected from WASH committee, health extension workers, and *Woreda* Edja Woreda Water, Mineral and Energy Office officials through interview to include expert's idea.

3.7 Instrumentation

The specific instrument for the primary data collection was through questionnaire survey and interview from *Woreda* government staffs and users of potable water services in purposively selected *Kebeles* in the *Woreda.* Secondary data was also collected through document revision in government and NGO offices. Physical observation of water facilities in the study areas and informal interview was also made to different individuals to generate additional information.

3.7.1 Survey questionnaire

Questionnaire was developed to address the set objective for the research project. The questionnaire was translated in to local language (Amharic) for the easy of understanding of it by the respondents. The questionnaire was distributed to Edja *Woreda* Water, Mineral and Energy Office officials, and health extension workers to include expert's idea. The questionnaire was also distributed to the beneficiaries of potable water in the selected *Kebeles* of the *Woreda.* Accordingly, the filled questionnaire was be checked and coded for consistency.

3.7.2 Interview

Structure interview was prepared to generate additional information for the research project. Thus, the interview was made with Edja *Woreda* Water, Mineral and Energy Office officials, and focal persons of water actors in the study area. Accordingly, four experts from government and two officials from NGOs office were interviewed.

3.7.3 Focus Group Discussion (FGD)

The research project was supplemented by the information generated from focus group discussion. Thus, discussion was made with a group comprises eight individuals in the Desene *Kebele*. The discussion was made at the favorable place and time for the participants. Short hand note was used to record data during the focus group discussion.

3.8 Sampling methods

From the total *Kebele* in the *Woreda*, six sample *Kebeles* was purposively selected and one person whose age was greater than 18 was invited to fill the questionnaire from randomly selected households to generate water management issues. From these six *Kebeles*, a total of 265 individuals was contacted and fill the questionnaire through the assistance of enumerators. The population sample in each *Kebele* was selected by proportionate stratified random sampling. The stratum was selected *Kebeles* in the *Woreda*, where water scheme is available to identify its challenge and prospects. The total population covered by this water system was estimated to be 25% of the total *Kebele* population from which 5% of the water beneficiaries were covered by the study. But, WaSH committee, and government water experts was contacted purposively based on their responsibility in regard to potable water and the knowledge they have about the potable water services in the *Woreda*. The sample frame and the sample population of the study are both represented in table 2 as follows.

Table 2. Sample frame of the sample *Kebeles*

S/N	Sample *Kebele*	*Total population	25% of the total *Kebele* popula-tion	5% of the 25% of the *Kebele* population (proportionate stratified random sampling)
1	Sheramo	4559	1140	57
2	Shebreden	2845	711	36
3	Desene	2742	686	34
4	Agena 01	4281	1070	53
5	Weyra Debane	4259	1064	54
6	Wasamar	2531	632	31
Total sample size				**265 respondents**

*Source: Edja Woreda Finance and Economic Development Bureau, 2011

3.9 Methods of data analysis

After all necessary data was collected; the questionnaire was checked and the qualitative data was coded quantitatively to simplify the process of qualitative data analysis. Accordingly, the entire collected and coded questionnaire was fed in the data view of the SPSS. Thus, the SPSS version 13 was used to analysis raw data using descriptive statistics such as; frequency, cross tabulation and percentage distribution of the population. The Excel 2007 was also used to produce graphing of frequency distributions. The unit of analysis of the study was groups of potable water actors. Hence, the responsiveness, accountability, and the networking mechanism among actors of potable water in the study area were analyzed to describe the challenge and prospects of potable water management in the *Woreda.* The economical, institutional and social structures of potable water actors were also assessed to describe the challenge and prospects as well as the sustainability of potable water services in the study area.

3.10 Ethical consideration

Ethics is inseparable from political, social and economical aspect of the research. Thus, great consideration was given to it during the study process. Ethics such as respect for persons, informing purpose of the study and research consent, creating confidence for the respondents, expressing in good manner, taking in to account the expression of truthfulness, avoid cheating and data falsification was practiced in the study especially during data collection.

3.11 Summary of Methodology

The study used mixed social science research method that uses qualitative and quantitative data. The main data collection method was survey questionnaire, interview, focus group discussion and physical observation. Six total *kebeles* were selected in the Edja Woreda and proportionate stratified random sampling method was employed to select 265 respondents in those six *kebeles*.

After all necessary data were collected; the questionnaires were coded and processed to generate descriptive statistics such as: frequency, tabulation, and graphing and percentage distribution using the SPSS software version 13. Thus, interpretation were made from the processed data to answer the research questions mentioned under chapter one.

CHAPTER FOUR

RESULTS AND DISCUSSION

This chapter presents analysis of the data collected using survey questionnaire, interview, focus group discussion and physical observations. Thus, the chapter try its best to answers the research questions mentioned under chapter one. Hence, the role of water management actors, the networking status among them, their responsiveness and accountability, the water supply capacity and distributive mechanisms and the sustainability instruments were discussed under this chapter.

4.1 Background characteristics of the survey respondents

This study was dominantly used primary data which was collected from the beneficiaries of potable water in Edja *Wored*a of Guraghe Zone through questionnaires. Thus, the majority of the study finding reflects the water user's view. Hence, it is found to be helpful by the researcher to describe background characteristics of the respondents like; age, level of education, occupation and family numbers to give the social image of these respondents to the reader. Besides, knowing the family size of the respondent can help the reader to easily understand the findings of water supply capacity in the study area.

A total of two hundred sixty five potable water beneficiaries were contacted during the data collection stage of the study from which primary data was collected through questionnaire. From the 265 respondents, 144 were male while the other 121 were female that comprises 54.3 and 45.7 percent of the contacted individuals respectively. Regarding their marital status, the majority of the respondents, 122 (46%), were single, 118 (44.5%) were married, 16 (6%) were those who lost their marriage partner by death and the remaining 8 (3%) were divorced. Among the contacted respondents, 88 or 33.21% of the respondents have 4-5 family members while 68 (25.66%) of them have 2-3 family members. The other 67 (25.28%) have 6-7 family members. The left 27 (10.19%), 4 (1.51%) and 2 (0.75%) have 8-10, 11-14 and 16 and above family members respectively. A considerable number of the respondents who participated in this survey were illiterate (21.13%) of the total respondents.

As it is indicated in the table 3 bellow, 51(19.25%) of the respondents have 9-10 grade level of education while 31 (11.70%) of the respondent have primary level of education. The other 22(8.3%) and 42 (15.85%) have 11-12 and college level of education respectively.

Table 3. Family size and the educational status of the respondent

S/N	Family size of the respondent			Educational status of the respondent		
	Family Number	**Frequency**	**Percent**	**Educational Status**	**Frequency**	**Percent**
1	No response	9	3.40	No response	7	2.64
2	2-3	68	25.66	illiterate	56	21.13
3	4-5	88	33.21	Grade 1-5	31	11.70
4	6-7	67	25.28	Grade 6-8	56	21.13
5	8-10	27	10.19	Grade 9-10	51	19.25
6	11-14	4	1.51	Grade 11-12	22	8.30
7	16 and above	2	0.75	College Graduate	42	15.85
	Total	**265**	**100.00**	**Total**	**265**	**100.00**

Source: Own data computation, 2012

Most of the respondents are within the age limit of 18-25 that accounts 32.5% of the total respondent followed by 36-40 age limits that have 19.6% of the total. The age of the respondents are also represented by figure 2 below.

Figure 2: Ages of the respondent

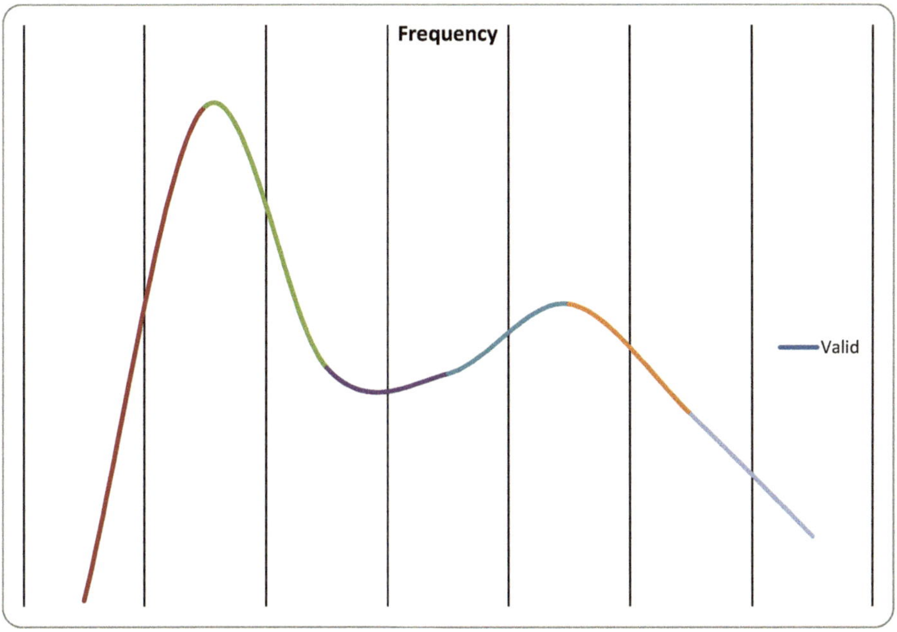

Source: Own data computation, 2012

As it is indicated in the table 4 bellow, most of the respondent of the study are farmers which is 32.45% of the total respondents while 25.7% of them are students. 47 (17.74%) of the respondents are government employees followed by merchant, artisan and jobless that have 10.94%, 6.79%and 4.15% respectively.

The detail explanation of age and occupation of the respondent is described as follows by table 4 below.

Table 4. Age and Occupation of the respondent

S/N	Age of the respondent			Occupation of the respondent		
	Age	**Frequency**	**Percent**	**Occupation**	**Frequency**	**Percent**
1	No response	1	0.38	No response	2	0.75
2	18-25	86	32.45	Farmer	90	33.96
3	26-30	41	15.47	Merchant	29	10.94
4	31-35	40	15.09	Artisan	18	6.79
5	36-40	52	19.62	Government employee	47	17.74
6	41-50	33	12.45	Student	68	25.66
7	51 and above	12	4.53	Jobless	11	4.15
	Total	**265**	**100**	**Total**	**265**	**100**

Source: Own data computation, 2012

4.2 The role of potable water management actors

The major potable water actors in the study area according to the interview conducted with potable water users are; Edja *Woreda* Water, Mineral and Energy Office, NGOs (World Vision Ethiopia and Kalehiwot Church), CBO (WaSH committee). The smaller numbers of the interviewees (25%) have only known government and world Vision Ethiopia (WVE). The role of each water actors in the Edja Woreda of Guraghe Zone is discussed in the next part of this paper.

Table 5. Potable water actors in the study area

S/N	Potable water actor	Frequency	Percent
1	Government (WMEO), World Vision, Kalehiwot, and WaSH committee	6	75
2	WVE and Government	2	25
	Total	**8**	**100**

Source: Own data computation, 2012

4.2.1 The roles and responsibility of potable water actors

Based on the interview conducted with government officials, the type of water provision modality of the study area is public provision. The local government directly provides water services to the users with the help of the actors in water system development. The interview result also shows that the *Woreda* water, Mineral and Energy Office is one of the local government organizations that coordinate the entire effort made to improve potable water services to the community. Thus, the office is responsible to identify those areas in the community with serious potable water services problems and give special attention to those areas. The office is also fully responsible to coordinate and conduct studies that aim to identify potential springs and underground water that might be developed to serve the community.

Table 6. The role of Edja Woreda Water, Mineral, Energy Office

S/N	Role	Frequency	Percent
1	Supply experts	3	37.5
2	Coordinate overall potable water activities	5	62.5
	Total	**8**	**100**

Source: Own data computation, 2012

The Woreda Water, Mineral and Energy Office also work closely with the Guraghe Zone Water, Mineral and Energy Department to mobilize zonal experts on those water activities that require technical knowledge which is beyond the capacity of *Woreda* officials. Moreover, the *Woreda* Water, Mineral and Energy Office coordinate different potable water awareness trainings to the community. Based on the survey conducted, 43.02% of the respondent said that the *Woreda* Water, Mineral and Energy Office is also responsible for potable water scheme management in the *Woreda* (See table 9 below).

Furthermore, the *Woreda* Water, Mineral and Energy Office work closely with zonal and *Woreda* health offices to conduct biological and chemical test for the newly constructed water scheme in the *Woreda* beforehand. The office also plays guiding and mobilization role for potable water maintenance experts in case of water scheme damage to different areas in the *Woreda.* Generally, the *Woreda* Water, Mineral and Energy Office plays the overall coordination role in the development and management of potable water scheme as well as potable water services in the *Woreda.*

Non Governmental organizations (NGOs) are also one of the active potable water actors in the *Woreda* that work within the scope of government water development plan and policies. The major role of NGOs in the *Woreda* is financial support for spring development, spring capping, pipe extension, deep and shallow well drilling. As indicated in table 7 below, 75% of the interviewees said the major role of NGOs in the area is conducting potable water development activities.

Table 7. The role of NGOs

S/N	Role	Frequency	Percent
1	Spring development, deep and shallow well drilling, pipe extension	6	75
2	Financial support	2	25
	Total	**8**	**100**

Source: Own data computation, 2012

NGOs also finance and facilitate different awareness creation trainings to the selected members of the community in the area. Secondary data review also shows that World Bank and UNICEF support government budget for potable water development in the *Woreda* (ITAB Consult, 2011). But the separate water budget amount of different actors was not available at *Woreda* level.

Water Sanitation and Hygiene Committee (WaSH committee) is the community based organization that actively act on the area of potable water service activities. The Focus Group Discussion (FGD) result shows that WaSH committee work closely with the *Kebele* administration, *Woreda* Water, Mineral and Energy Offices and with NGOs to bridge information to and from the community. The committee also plays vital role in mobilizing the community during water scheme constructions. However, the major role and responsibility of the water committee according to the interview made with government official is post construction scheme management.

Table 8. The role of Wash committee

S/N	Role	Frequency	Percent
1	Post construction scheme management	6	75
2	Collect water fee	2	25
	Total	**8**	**100**

Source: Own data computation, 2012

The WaSH committees also manage potable water schemes on behalf of the community which is crucial for the sustainability of water scheme. They committee collects water fees from the beneficiaries and use it for the maintenance purpose in case of water scheme damage. In addition to its managerial role, the WaSH committee also plays monitoring and evaluation role on water scheme and potable water services. Thus, the WaSH committees supervise the potable water services and arrange water service shift in consultation with *Woreda* Water, Mineral and Energy Office. The *Kebele* administration office also plays coordination role at grass root level.

According to the finding of this survey, 122 out of 265 or 46.04% respondents said that WaSH committee is responsible for the water scheme management in the study area.

Table 9. Responsible body of potable water scheme management

S/N	Potable water actor	Frequency	Percent
1	No response	14	5.28
2	Woreda water, Mineral and Energy Office	114	43.02
3	WaSH Committee	122	46.04
4	*Kebele* administration	12	4.53
5	Woreda administration	3	1.13
	Total	**265**	**100.00**

Source: Own data collection, 2012

Based on the survey result, the potable water beneficiaries are active participant in the construction process of water schemes in the study area. The beneficiary participation on the water resource development from the view of the respondents is given in table 10 as follows.

Table 10. Beneficiary participation in potable water resource development

S/N	User participation	Frequency	Percent
1	No answer	21	7.92
2	Participated	215	81.13
3	Never participated	24	10.94
	Total	**265**	**100.00**

Source: Own data computation, 2012

As it is indicated in table 10, 215 or 81.13% of the respondents replied that they have participated in the activities of potable water scheme construction. From this total, 44.9% of them have been participated by labor supply while 39.6% of them have participated by financial support. Among the contacted respondents, 50.2% of them said that the beneficiaries of potable water are responsible to penalize someone who damage water schemes deliberately.

Table 11. Means of participation on potable water resource development

S/N	Means of participation	Frequency	Percent
1	No answer	36	13.58
2	Financial support	105	39.62
3	Material support	4	1.51
4	Labor supply	119	44.91
5	Professional assistant	1	0.38
	Total	**265**	**100.00**

Source: Own data computation, 2012

In general speaking, the potable water actors are playing vital role to manage and sustain the potable water services in the woreda. But the management role is almost left to government bodies except WaSH committee. Emperical studies reveals that developing countries have already adimited that they cannot solve their deep rooted potable water problems working alone and their governance system must permit all stakeholders to actively engage and solve the growing potable water problems. More effective management systems need to be created to overcome government failure, market failure and system failure or a combination of these (Rogers and W.Hall, 2003). Thus, potable water and secheme management should not left only to the government and WaSH Committee. All actors, including the beneficieries and NGOs should take part in the post construction managemnt, monitoring and follow up of the water seystem. Post construction secheme monitoring is the the vital activity to ensure sustainability but the respondets have no idea about the specific water actor that is responsible for monitoring. This implies the absence of continous monitoring activity on the potable water scheme.

4.2.2 Status of responsiveness of potable water actors

Regarding the responsiveness of the potable water actors, majority of the respondents (64.15%) of them answered that the Woreda Water, Mineral and Energy Office of the area is responsive to the beneficiaries. On the other hand, 30.6% percent of the respondent said that the office is not responsive. They reason out that the government sector offices lack responsiveness since it neither works closely with the beneficiaries to identify their need nor actively respond to peoples demand on time. Regarding NGOs, 75.85% of the respondents replied that World Vision is responsive while 16.23 said World Vision is not responsive to the demand of the community. On the other hand, 93% of the respondent confirmed that Kalehiwot is responsive but the significant number or 50.94% didn't respond about Kalehiwot since it is operating in other parts of the Woreda. The study also assessed the responsiveness of the WaSH committee. Accordingly, 69.43% of the respondents indicated that the WaSH committee is responsive. But 15.85% of them pointed out that the WaSH committee is not responsive to the community. Similarly, the focus group discussion result shows that potable water actors are responsive but there is delay in their response. Based on the interview made with government officials, water actors of the area are not responsive to the beneficiaries of potable water since there is financial and human resource capacity limit.

Table 12. Responsiveness of major potable water actors to the beneficiaries

		Responsive		**Not Responsive**		**No answer**	
S/ N	**Major potable water Actors**	**Frequency**	**Per-cent**	**Frequency**	**Percent**	**Frequency**	**Percent**
1	Water, Mineral and Energy Office	170	64.15	81	30.57	14	5.28
2	World Vision Ethiopia	201	75.85	43	16.23	21	7.92
3	Kalehiwot	93	35.09	37	13.96	135	50.94
4	WaSH Committee	184	69.43	42	15.85	39	14.72

Source: Own data computation, 2012

4.2.3 Status of accountability of potable water actors

Most of the respondents answered that Water, Mineral and Energy Office is fully accountable to beneficiaries. Hence, 78.5% of the respondents reported that the office is accountable to their action and the decision. Nevertheless, 19.6% of the respondents have complained that the office is not accountable for their action. On the other hand, 84.15% of the respondents revealed that World Vision is also fully accountable, even though, 12.83% of the respondents rated as unaccountable. The majority of the respondents (68.68%) were not well aware about the accountability of Kalehiwot. Thus, only 21.89% of the respondents rated Kalehiwot as accountable while 9.43 said unaccountable. Regarding the WaSH committee, 70.94% of the respondents rated them as accountable while 27.92% said WaSH committee is unaccountable.

Table 13. Accountability of major potable water actors

S/N	Major potable water Actors	Accountable		Not accountable		No answer	
		Frequency	Percent	Frequency	Percent	Fre-quency	Percent
1	Water, Mineral and Energy Office	208	78.49	52	19.62	5	1.89
2	World Vision Ethiopia	223	84.15	34	12.83	8	3.02
3	Kalehiwot	58	21.89	25	9.43	182	68.68
4	WaSH Committee	188	70.94	74	27.92	3	1.13

Source: Own data computation, 2012

As discussed in the preceding parts of the paper, potable water actors are both responsive and accountable to their actions. But it needs more improvement to establish good potable water management. For IFAD, the term good governance which is characterized by different elements including responsiveness and accountability is highly correlated with the notion of sustainable development (IFAD, 1999). The potable water actors in the area should increase their degree of responsiveness and accountability to ensure the sustainability of potable water services in the *Woreda* since significant numbers of the respondents are still saying they are not both responsive and accountable. This shows that there is a governance gap from the side of water actors. In the society centric approach of governance, alternative modes are less reliant on top down policy instrument but need to find out more accountable and democratic means of social organization where responsibility and accountability for intervention is shared equally between public, community and private actors (Allen et al., 2006). Therefore, frequent community involvement is needed to share responsibility and accountability.

4.3 The networking status among major potable water actors

The interview result from *Woreda* Water, Mineral and Energy Office officials reveals that there is good networking status among major potable water actors in the *Woreda.* However, the results from the survey conducted up on the potable water beneficiaries shows a slight different result. Among the total respondents, 24.5% of them said that the networking status among water actors is poor or very poor while 67.2% of the respondents rated the networking status among potable water actors as; good or excellent.

The interview results of government officials and focus group discussion indicates that the networking status of water actors in the area is good. ITAB consult also concluded that the networking among potable water actors is good in the *Woreda* (ITAB Consult, 2011). Good networking increases the effort to achieve potable water supply, increase responsiveness and avoid duplication of efforts. The network management theory states that a governance network is self-regulating networks that contribute the production of public purpose. It is a particular form of governance which includes a relatively stable horizontal articulation of interdependent, but operationally autonomous actors (Sørensen and Torfing, 2005). But the study result reveals that there is still networking gap among actors since 24.5% of the respondents said the network among actor is poor or very poor. Thus, for achievement of good potable water services in the area, the existing network among actors should be improved as much as possible. There should be well stable networking between water actors that water related information flow easily and the cooperation between the community and actors will get strong. Whenever there is good networking, the existing water service related problem will not exist. Therefore, the study result reveals that the network among water actors in the area is not to the extent that promotes good water management.

Table 14. Level of networking between potable water actors in the *Woreda*

S/N	Status	Frequency	Percent	Cumulative Percent
1	No response	4	1.51	1.51
2	Very poor	17	6.42	7.92
3	Poor	48	18.11	26.04
4	Good	75	28.30	54.34
5	Very Good	89	33.58	87.92
6	Excellent	14	5.28	93.21
7	Have no knowledge	18	6.79	100.00
	Total	**265**	**100.00**	

Source: Own data computation, 2012

4.4 The status of potable water management in the *Woreda*

Every natural resource can shorten below the needed level unless managed properly. On the other hand, better resource management can improve the efficiency of its usage and prolong its life time. For the better management of the natural resources like potable water, it needs to identify those challenges that inhabit better potable water services in the study area.

4.4.1 Supply capacity

Based on the interview result from government and NGO officials, the study areas of the Edja *Woreda* have one water supply system, which is Fessa Water Supply that has the supply capacity of 30 liter/second and covers 32 km including the sample *Kebeles.* Even though the sample Kebeles have same water supply system, there is variation in the amount of the water that the respondents use and can get per day per household. Those who live in the upper course of the water system can get much more water than those who live in the lower course of the water system. The majority of the respondents 21.89% said that they use 36-45 liter per day per family.

The other 19.25% indicated that they use 26-30 liter per day for their entire family water consumption. The other 18.49% and 15.47% uses 46-75 and 10-20 liters per day per household respectively.

The average amount of water that the respondents can get per day per household was also assessed in order to address the supply capacity from the respondents' view. Accordingly, the higher percentage of the respondents (16.98%) said that they can get 10-20 liter per day per household. The other 15.85% said that they can get 75-100 liter of water per day per household. These figures imply that there is significantly higher variation in supply capacity between the higher and lower course of the water system. Thus, the water supply system is suffering from equity problem due to the lack of wise and well programmed water usage. This is one of the potable water management challenges of the study area. Table 15 below illustrates the amount of water the respondent use and can get per day per household.

The study result also implies that the water supply capacity of the water system in the area is very low which is far below the Millennium Development Goals (MDGs) target which is 100 liter per day per household for an average family size of 5 persons (Mwebaza, 2010) and the minimum definition of basic water supply. The basic potable water supply is defined as the provision of effective water use as well as a minimum quantity of 125 liters of potable water per household per day (or 6000 liters per household per month) within 200 meters of a household and with a minimum flow of 10 liters per minute in the case of communal water points (Khambule, 2002). Thus, the water supply capacity needs dramatic improvement to meet the definition of basic potable water supply and to meet the MDGs target.

Table 15. The amount of potable water the respondent use and can get per day per household

S/N	The amount of water respondents use per day per family			The amount of water respondents can get per day per family		
	Volume in Liter	**Frequency**	**Percent**	**Volume in liter**	**Frequency**	**Percent**
1	No response	3	1.13	No response	10	3.77
2	5-10	17	6.42	10-20	45	16.98
3	11-15	8	3.02	11-20	1	0.38
4	10-20	41	15.47	21-30	39	14.72
5	21-25	17	6.42	31-40	27	10.19
6	26-30	51	19.25	41-50	31	11.70
7	31-35	13	4.91	51-60	17	6.42
8	36-45	58	21.89	61-70	12	4.53
9	46-75	49	18.49	75-100	42	15.85
10	76-100	8	3.02	101 and above	35	13.21
	Total	**265**	**100.00**	As much as needed	6	2.26
				Total	**265**	**100.00**

Source: Own data computation, 2012

4.4.2 Distributive Mechanisms

Most of the respondents that reside around the upper course of the water system revealed that the water distribution in their area has on good status and capacity. But the majority of the contacted respondents, 126 (47.5%) of them answered that the water is not available most of the time. But, a significant number of the respondents didn't answer the question since they are satisfied with the existing potable water services. On the other hand, 49.1% of the respondent indicated that the water scheme they use have got distributive problem like, less water power, pipe leakage and unavailability of the water when needed. Problems like less water power and water interruption are especially observed in the lower course of the water line.

Table 16. The major challenges of potable water management

S/N	Distribution problems	Frequency	Percent
1	No response	71	26.79
2	The water is not available most of the time	126	47.55
3	Long distance of water point	3	1.13
4	Lack of good monitoring system in place	2	0.75
5	The water have the problem of impurity	3	1.13
6	Lack of sanitation around water point	3	1.89
7	Lack of efficient water utilization plan	29	10.94
8	There is no responsible body to coordinate people	9	3.40
9	Water users do not practiced what they trained	2	0.75
10	Lack of finance for water point	1	0.38
11	Disagreement among water users	1	0.38
12	Pipe leakage/breakdown	5	1.89
13	Lack of equity in water usage	3	1.13
	Total	**265**	**100.00**

Source: Own data computation, 2012

The focus group also indicated that the water is not available most of the time especially in the dry season. They also pointed out that the water is available every time in some *Kebeles* of the *Woreda* while it is hardily available in the lower course of the water system which shows inefficient utilization and poor management. On the other hand, the interview conducted with government officials reveals that the water power decrease during dry season. Besides, the water has recently extended to other areas that it couldn't reach every beneficiaries at a time.

Even though significant numbers of the respondent didn't comment on the water distribution capacity of the area, the study result reveals that water distribution has get serious problem in the area. The focus group indicated that some areas of the water system suffer from shortage while there are overflows of water in the other area. This also strengthens the fact of the absence of water usage program in the area.

The study result implies that the water supply is interrupted most of the time which is the major distribution problem in the area. On the other hand, the basic potable water supply is defined as the provision of effective water use as well as a minimum quantity of 125 liters of potable water per household per day which is not interrupted for more than seven days in any year (Khambule, 2002). Thus, the facts revealed in the study imply that water distributive mechanism is not functioning to the extent that fulfills the basic potable water supply in the study area.

4.4.3 Distance

The distance of the potable water point that the community uses for drinking purpose is another accessibility criterion for potable water (Ibid). The study finding reveals that the distance of water point is one of the water management challenges in the study area. The focus group answered that they can get water in nearby area for themselves. But there are many individuals that still travel long distance to collect drinking water since there is no water system in their area. The time that water users take to collect potable water in the study area according to the respondents' points of view is summarized by table 17 below.

Table 17. The time traveled to collect potable water

S/N	Minutes travelled to get potable water	Frequency	Percent
1	No response	8	3.02
2	1-5	86	32.45
3	6-10	58	21.89
4	11-15	35	13.21
5	16-20	41	15.47
6	21-25	7	2.64
7	26-35	12	4.53
8	36 and above	3	1.13
9	0	15	5.66
	Total	**265**	**100**

Source: Own data computation, 2012

The higher percentage of the respondents or 86 (32.45%) of them traveled only 1-5 minutes to collect water for consumption purpose. The other 21.45% travel 6-10minutes to get water. This result shows that more than half of the respondents travel less than or equal to ten minutes to collect water. Among the total respondents, 29.43% travel more than 15 minutes to collect their potable water.

Figure 3. The time users take to collect potable water

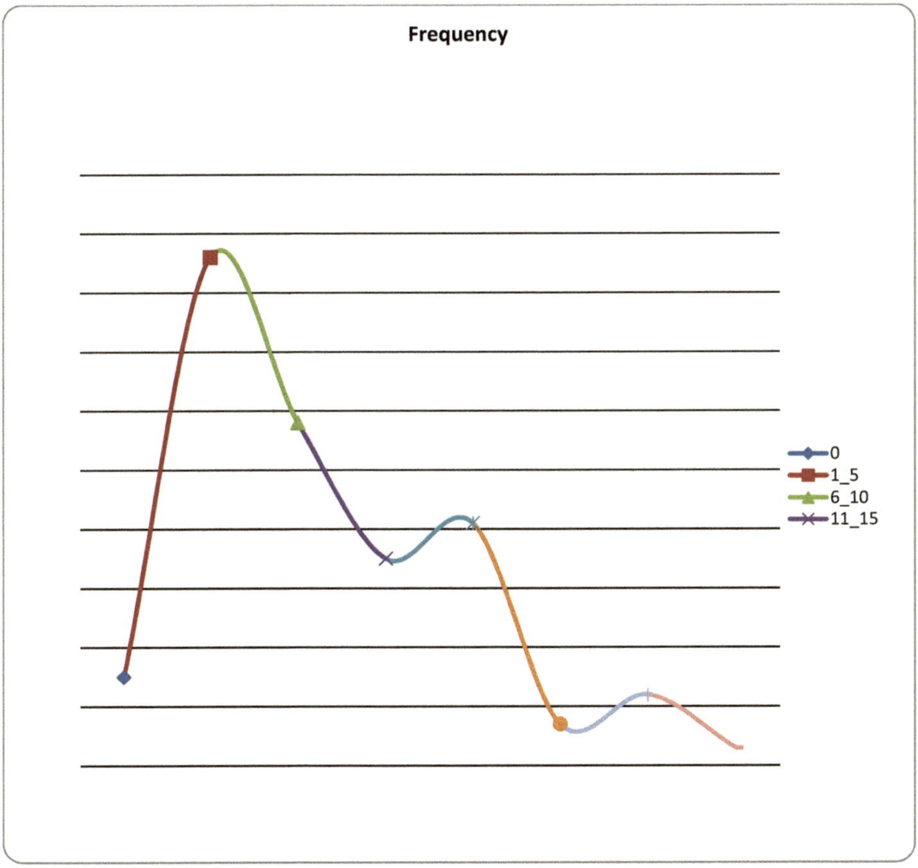

Source: Own data computation, 2012

The study result implies that the distance that majority of the water users' travels are less than 10 minutes. However, 42.64% of the respondents travel more than 10 minutes which is out of the definition of basic potable water defined under section 4.4.1 which states that the distance should not be more than 200 meters.

4.5 The challenges and prospects towards sustainability

The existence of institutional, economical and social mechanism determines the sustainability of one development project. Thus, the statuses of these sustainability variables have been assessed from the potable water beneficiaries, focus group discussion and water actors' officials' point of view.

4.5.1 Institutional mechanisms

Institutional mechanism in this study refers water regulatory bodies, management arrangement, and planning procedures that determines the sustainability of development projects. Most of the respondents said that WaSH committee is responsible for the water scheme management in their area. Accordingly, 45.7% of the respondents indicated that WaSH committee has been playing both the management and regulatory role of water scheme in their area. On the other hand, 41.9% of the respondents answered that the Edja *Woreda* Water, Mineral and Energy Office plays the management of water scheme. Here one can understand that the respondents are not clear about whether WaSH committee is fully responsible for the management of water scheme or *Woreda* Water, Mineral and Energy Office. But based on the interview made with *Woreda* Water, Mineral and Energy Office officials, WaSH committee is fully responsible for the management of water scheme in the area. The focus group discussion shows that the WaSH committees in consultation with *Woreda* sector office manage the water scheme of their area.

According to the survey result, 78.49% of the respondents indicated that there is clear procedure for the planning of potable water project in their area. In contract to this, 10.19% of the respondent answered that there is no clearly indicated planning procedure for potable water projects for the reason they do not have enough information. In general speaking, there is good prospect for the institutional mechanisms in place to foster the sustainability of water services. But there are the challenges of clearly specified management arrangement and regulatory bodies.

Table 18. Availability of clear planning procedures

S/N	Availability planning procedures	Frequency	Percent
1	No answer	27	10.19
2	There is clear planning procedures	208	78.49
3	There is no clear planning procedures	28	11.32
	Total	**265**	**100.00**

Source: Own data computation, 2012

when a unit of common pool resources have high value with the absence of institutional restrict to limit the way resource units are appropriated, individuals get strong incentives to appropriate more and more resource units leading to congestion, overuse, and even the destruction of the resource itself (Ostrom, 1999). Thus, the study result implies that the institutional mechanism in place is not strong enough to limit and regulate the destruction of potable water scheme in the area. The study conducted by ADB also revealed that the weak institutional mechanisms in Ethiopia have limited the planning; implementation and monitoring capacity at all level (ADB, 2005). Harries et al. also added that the local institutions are often weak to ensure potable water services sustainability (Harris et al., 2011).

4.5.2 Economical mechanisms

Economical mechanism in this study refers measures used to encourage efficient utilization of potable water resource. Hence it comprises water pricing, charging and water related penalties. Most of the water users that participated in the study reported that they are paying water fee. Accordingly, 218 or 82.3% of the respondents answered that they are paying water fee for the water they are using. Among the respondents, 54.3% reported that the water fee collected from potable water users is used for maintaining existing water scheme in case of damage.

Table 19. Water fee payment status from the view of the respondent

S/N	Water fee payment	Frequency	Percent
1	No answer	25	9.43
2	Pays water fee	218	82.26
3	Not pay water fee	22	8.30
	Total	265	100.00

Source: Own data computation, 2012

According to the focus group discussion results, the water fee in the study area is not based on the amount of water that they use which is inefficient since there is no mechanism to control the amount of water consumed by a given household. Thus, simply fixed amount of money is imposed on the given household which is to be paid monthly.

The focus group also revealed that the drinking water fee that users paying is used for the maintenance of water scheme. But according to the physical observation made and interviews conducted with NGO officials, the maintenance of water scheme is not in place and there were even a case when NGOs participated in maintenance activities. For ADB, sustainability will be assured by making the beneficiary communities fully responsible for the operation and maintenance costs as well as the management role (ADB, 2005). Despite these facts, the secondary data review pointed out that there were problems in community's water point's management. "Pipe fittings were damaged; water point bodies and fences were demolished due to poor management by the community. Thus, World Vision Ethiopia in collaboration with the community of the area and the local government has maintained potable water schemes to maximize the efficiency of the project" (WVE, 2012 P.1). This fact indicates that there are no well designed economic mechanisms to encourage efficient utilization of potable water resource in the study area.

Among the contacted respondents, 204 (77%) of them answered that penalty system is in place that anyone who damage water scheme is penalized. However, the respondents unable to clearly identify the responsible body that penalizes misuse of water resources or water schemes in their area.

Table 20. Payment of water fee and the availability of water misuse penalty system

S/N	Do water users pay potable water fee			Availability of responsible body to penalize		
	Water fee Payment	Frequency	Percent	Responsible body	Frequency	Percent
1	No answer	25	9.43	No answer	26	9.81
2	Pays water fee	218	82.26	There is responsible body	204	76.98
3	Not pay water fee	22	8.30	There is no responsible body	35	13.21
	Total	**265**	**100.00**	**Total**	**265**	**100.00**

Source: Own data Computation, 2012

50.2% of the respondents answered that the potable water users or the community itself is responsible to penalize water related crimes.

The other 7.92% of the respondents said that WaSH committee and the *Woreda* Water, Mineral and Energy Office are responsible to penalize. The focus group said that even if WaSH committee and *Woreda* sector office is responsible to penalize, penalty didn't yet come in to practice in the area. But based on the interview made with *Woreda* water, Mineral and Energy Office Officials, Edja *Worda* Water, Mineral and Energy office is fully responsible to penalize the misuse of water resources and potable water schemes.

The physical observation of the water scheme in the study area shows the practical functioning of the Garrett Hardin's tragedy of commons (Hardin, 1968). There is leakage and worn out of water schemes and every individual think only about the water they get without taking in consideration about the damage and maintenance of the water scheme in the study area.

Table 21. Responsible body for penalizing misuse or damage

S/N	Potable water actor	Frequency	Percent
1	No response	87	32.83
2	The potable water beneficiaries	133	50.19
3	WaSH committee	21	7.92
4	The beneficiaries and the WaSH committee	3	1.13
5	Woreda Water, Mineral and Energy Office	18	6.79
6	No assigned body to punish	3	1.13
	Total	**265**	**100**

Source: Own data computation, 2012

The above facts reveal that there are good prospects of economic mechanism in that the mechanisms are in place. The challenge is that the economic mechanism is hardly working to ensure the sustainability of water services in the area. On the other hand, the respondents have no clear idea who penalizes deliberate scheme damage since the responsible body didn't practice it yet. On the other hand, water fee is unable to maintain damaged scheme which shows lack of future service sustainability.

4.5.3 Social mechanisms

Social mechanisms indicate those measures to increase awareness of water issues and mobilize users to participate in planning, management and financing of water resource development. One of the common measures to increase awareness of water issues in the study area is through training. Thus, the respondents were asked whether they have undertaken water related trainings. Accordingly, 43.8% have answered they have already taken water related trainings. But majority (51.3%) of the respondents indicated that they did not take any water related trainings. However, they have heard water related issues through public gathering and discussion. Most of the respondents are well aware about water management that Water, Sanitation and Hygiene committee who are selected community members are established in each *Kebele* of the study area.

Table 22. Water related training given for the potable water users

S/N	Water related training	Frequency	Percent
1	No answer	13	4.91
2	Take water related training	116	43.77
3	Didn't take water related training	136	51.32
	Total	**265**	**100.00**

Source: Own data computation, 2012

Regarding users participation in water scheme construction, 81.1% of the respondents answered that they have participated in water resource development projects. Among the 81.1%, 105 of them reports that they have participated in financial support, while 119 of them have participated in providing labor supply to drinking water resource development activities. The remaining respondents participated by provision of locally available construction materials and by professional support. The focus group discussion result also reveals that the potable water users participate in drinking water development in those ways like labor supply, and material support.

Thus, the study result shows that there is good social mechanism in place without which user participation could not have been happened. A significant numbers of the respondents have also already had awareness about water issues which is the promising prospect. ITAB consult also indicated that there is interesting community participation in water resource development. But basic awareness about potable water is low among the community of Edja *Woreda* for the practical implementation (ITAB Consult, 2011). The main challenge lies on the fact that the majority of the respondents didn't participate on the planning stages of drinking water development. The water resource management and water services functions more effectively with an open social structure which enables broader community participation (Rogers and W Hall, 2003). The study result implies that awarness about water and paticipation of water users are aleady in place, but users paticipation in management and planning lags behing the status of effectiveness.

4.6 Suggested solutions for the enhancement of better potable water management

The study also assessed the possible solutions of the potable water management problems in the *Woreda* from the view point of the beneficiaries, government and NGO officials. Accordingly, 18% of the respondents recommended that every household would better have water point at home that can increase more sense of ownership and help them to pay reasonable water fee. The other 15% of the respondent suggested that water services in the area needs improvement. 9% of the respondents commented that water related awareness training is needed for the potable water users in the *Woreda.*

The other possible solutions raised by water users includes; establishments of good monitoring and maintenance system, convincing the beneficiaries to pay reasonable water fee, cleaning the reservoir timely and adding chlorine to it, and assigning clearly identified water scheme monitoring body. The interview made with government and NGO officials reveals that water awareness creation for the water users and good cooperation and networking among actors can solve water management problems in the *woreda*. The views of respondents suggested solutions are summarized in table 23.

The view of the respondents implies that there is a wide range of management gaps that needs to be addressed for the better potable water services. Majority of the respondents called for the especial attention of the water management. Post construction scheme monitoring, keeping water points dry and clean, reasonable water fee payment and better water services are among the suggested solution that needs especial attention. It also needs to work on institutional capacity building to enhance the better water services in the area.

Table 23. Suggested solution for better potable water management

S/N	Suggested solution	Frequency	Percent
1	No response	56	21
2	Water should be used wisely	5	2
3	Encourage people to use toilet	2	1
4	Try to give better water services	40	15
5	There should be good monitoring and maintenance system	21	8
6	The water users should pay reasonable water fee	4	2
7	Clean the reservoir on time and add chlorine to it	7	3
8	Community participation should increase on water project	20	8
9	Expand water services to every household	47	18
10	More training should be given to beneficiaries	24	9
11	Keep the water clean	9	3
12	All stakeholders would better work in coordinated way	13	5
13	Assign water scheme monitoring body	8	3
14	Prevent water related corruption	4	2
15	Keep water point areas dry and clean	5	2
	Total	**265**	**100**

Source: Own data computation, 2012

CHAPTER FIVE

SUMMARY, CONCLUSIONS AND RECOMMENDATIONS

5.1 Summary of results

This study was designed to find out the challenges and prospects of potable water management in the Edja *Woreda* of Gurahe Zone. Accordingly, efforts have been made to find out the role of water actors in the *Woreda,* the networking status among water actors, the responsiveness and accountability of potable water actors, the status of potable water management in the area in terms of supply capacity and distributive mechanisms and the challenge and prospective of institutional, economic and social mechanism in place to ensure the sustainability of potable water management in the study area.

The study result indicates that the major potable water actors of the *Woreda* are Edja *Woreda* Water, Mineral and Energy Offices, NGOs (World Vision Ethiopia, Kalehiwot Church), multilateral agencies (UNICEF and World Bank), Water, Sanitation and Hygiene (WaSH) Committees and the beneficiaries of potable water in the area. The *Woreda* Water, Mineral and Energy Office coordinate the entire potable water activities, trainings and water potential area studies. On the other hand, NGOs and multilateral agencies finance potable water resources development and water related capacity building activities in the *Woreda.* The WaSH committees manage, collect water fees, report and facilitate the maintenance of malfunctioning potable water schemes.

The majority of the respondents indicated that the potable water actors are both responsive and accountable to the users. However, a significant percentage of the respondents complained that water actors in that both responsiveness and accountability lacks from the side of actors.

The networking status among potable water actors of the area is good from both the view of the beneficiaries and government officials. Nevertheless, the respondents also indicated that the network status among potable water actors need to be improved for the better development of potable water services in their area.

The study result reveals that it needs more effort to improve the distribution mechanisms of potable water resource in their area. The water is not available most of the time when the users need it. Pipe leakage, water wastage due to the lack of wise usage and lack of well planned maintenance system are some of the major distribution challenges of potable water in the study area.

The study result also reveals that those beneficiaries who live around the upper course of the water scheme have better access to potable water than those who lives around the lower course of the water system. The amount of water that most of the water users can get per day is below the average Millennium Development Goals target and the basic water supply definition.

Furthermore, the study result shows that some promising trends are observed in the establishment of institutional, economical and social mechanisms that enhance the sustainability of potable water management in the study area. WaSH committee has been established in every *Kebele* of the *Woreda* and they have started to facilitate water related activities. The water users start to pay water fee for the water they are using. Majority of the water users have also been participating in potable water resource development and the beneficiaries are well aware about water related issues. However, the mechanisms lack effectiveness to ensure potable water service sustainability.

5.2 Conclusions

The study results show that potable water management role is left for government bodies and the WaSH committee. This implies that there is potable water management gap in the area since the beneficiaries who use the water and NGOs who financed the resource development are not playing the management role. They are not taking part in the scheme management and evaluation that would have contributed for water scheme sustainability and hence potable water service sustainability in the area. The potable water actors of the area have been delivering what is needed on the basis of demand and with better accountability. But the degree of responsiveness and accountability needs improvement. The networking status among water actors is also good but need more effort to achieve the best standard.

The potable supply capacity of the study area has great problems. Both the water that a given household is using and can get is yet far below the definition of basic potable water supply and MDGs standard. The potable water distributive mechanism has also got problems like; unavailability of most of the time and the problem of equity in water distribution. A significant numbers of the beneficiaries also travel more than 15 minutes distance to collect potable water. The institutional, economical and social mechanisms all exist in the study area but they are all lacks effectiveness to ensure the sustainability of potable water services in the area. The study result implies that the potable water management issues like post water scheme construction monitoring and follow up is neglected by the water actors of the study area.

5.3 Recommendations

The study result indicates that there are promising trends in the management of potable water scheme and services in the study area. However, it needs more effort to establish the best status of potable water management in the Edja *Woreda.* Based on the study findings, the following points were recommended for the best drinking water services in the *Woreda.*

- Government bodies as well as NGOs and the water users should increase their management activities and post construction water scheme monitoring.
- It needs more effort to establish sense of ownership among water users.
- Great attention should be given to the capacity buildings of *Woreda* Water, Mineral and Energy Offices and the WaSH committee in every *Kebele* of the *Woreda.*
- More attention should be given to establish good networking on the monitoring and follow up of post water scheme construction.
- The responsiveness and accountability of local government and other actors should be improved significantly to meet the needs of drinking water users in the area.
- Water usage should be in programmed way and the time schedule should be communicated to the beneficiaries.
- It needs to assign responsible body that actively involved in the regulation and management of water scheme in the area.

- The water fee and the efforts of the community should be able to finance the maintenance cost for their own benefit. Therefore, further effort should be made to establish strong economical mechanism.
- Especial attention should be given to potable water user's participation in the planning stage of potable water project planning and resource development to create more sense of ownership. The water users should also play vital role in scheme monitoring and management.
- The challenges and prospective of water related issues like sanitation and hygiene status was not addressed by this study. Since sanitation and hygiene are both inseparable from potable water, further study is highly recommended to identify the challenges and prospects of sanitation and hygiene situation in the *Woreda* that it can supplement this study to be used for policy formulation.

Reference

Abdellatif, A. M. (2003, May 30). *Good Governance and Its Relationship to Democracy and Economic Development.* Retrieved December 6, 2011, from http://www.pogar.org/publications/governance/aa/goodgov.pdf

ADB. (2005, June). *Rural Water Supply and Sanitation Programme.*Tunis: ADP. Retrieved December 14, 2011, from http://www.afdb.org/fileadmin/uploads/afdb/Documents/Project-and-Operations/ET-2005-135-EN-ADF-BD-WP-REV.1-ETHIOPIA-AR-RWSSP-REV-1.PDF

ADP, WSP,AND UNDP. (2006, December). *Getting Africa on Track to Meet MDGs on Water and Sanitation.* Retrieved December 13, 2011, from http://www.wsp.org/wsp/sites/wsp.org/files/publications/319200725615_312007101903_MDGs_All_final3_high.pdf

Allen,A., D.Davila,J., and Hofmann,P.(2006). *Governance of Water and Sanitation Services for the Peri-urban Poor.* Retrieved December 9, 2011, from http://www.ucl.ac.uk/global-health/dpu1.pdf

Cde Zuene, J. (1997). *Handbook of Drinking Water Quality.* New York: John Wiley and Sons.Inc .

ECA. (2005). *Partnership Modalities for Enhancing Good Governance.* Addis Ababa: UNECA Development Policy Management Division.

Edja Woreda Finance and Economic Development Bureau. (2011). Edja Woreda Population by Kebele. 1.

GWP, EUWI and UNDP. (2008). *Economics in Sustainable Water Management.* Brassels: Cap-net publisher

Hardin, G. (1968). *The Tragedy of Commons.*New York: Jstor.

Harris, D.,Kooy, M., and Jones, L. (2011). *Analysing the governance and political economy of water and sanitation service delivery.* London: ODI.

IFAD. (1999). *Good Governance: An Overview.* Rome: IFAD.

ITAB Consult. (2011). *WaSH Baseline Survey For WVE Major Donor Ministry (MDM) WaSH Program.* Addis Ababa: WVE.

Izquierdo, E.L., Lucky, R., and Elvert, K.(2004). *Fresh Water Resources:Managing the Risks Facing the Private Sector.* California: Pcific Inistitute.

Kauffmann, C. (2007, April). *Access to Drinking Water.* Retrieved December 13, 2011, from http://www.oecd.org/dataoecd/27/2/38563239.pdf

Khambule,T. (2002, October). *Water is Life,Sanitation is Diginity.* Retrieved October 18, 2011, from www.africanwater.org/Documents/draft_SA_water_services_wp6.1.pdf

L. Moe, a. R. (2006). *Global challenges in water, sanitation and health*.Atlanta: Emory University.Retrieved December 14, 2011, from http://www.ipwr.org/documents/004S041.pdf

Melissa and Janson. (2008, January 01). *Home Sweet Home*. Retrieved October 15, 2011, from http://chsfsforums.websitetoolbox.com/post?id=2423469

Miranda, M. Hrdjidk, M. and Molina T.K., R.(2011, July). *Water Governance key Approaches: An Analytical Framework*. Delhi: Chance 2 Sustain.

MoFED. (2005). *A Plan for Accelerated and Sustained Development to End Poverty (PASDEP).* Addis Ababa: MoFED.

Mwebaza, R. (2010, January). *Sustaining good Governance in Water and Sanitation in Uganda*. Retrieved October 18, 2011, from http://www.iss.co.za/uploads/M168Full.pdf

Nigatu, L. (2011). *The Status of Children with Disability in Education and Health Services with Special Emphasis to Acess and Quality of the Services at Edja Woreda.* Addis Ababa: Unpublished.

Ostrom, E. (1999). *Coping with Tragedies of the Commons*.Bloomington: Indiana University.

ProAct. (2011). *The Major Donor Ministry (MDM) Water, Sanitation and Hygiene (WASH) Program Baseline Report.* Quacha Birra.

Rao, B. (2006). *Strategic Thinking to Achieve Water MDGs*. Retrieved 2011 09, December, from http://www.adb.org/Water/topics/mdgs/pdf/Strategic-Thinking-MDG.pdf

Rogers,P., and W.Hall, A.(2003, February). *Effective Water Governance*. Retrieved November 15, 2011, from http://www.orangesenqurak.com/UserFiles/File/GWP/GWP%20TEC%20Paper%207_English.pdf

Sørensen and Torfing. (2005). *The Democratic Anchorage of Governance Networks.* Roskilde: Nordic Political Science Association.

Toikka, A. (2011). *Governance Theory as a Framework for Emperical Research.* Helsinki: University of Helsinki.

UNDP and SNV. (2009, April). *Capacity Development for Water and Sanitation*. Retrieved December 14, 2011, from http://capacity.org/capacity/export/sites/capacity/documents/journal-pdfs/CAP36_0409_Watsan_ENG.pdf

UNDP,WGF and SIWI. (n.d). *Water and Sanitation Governance*. Retrieved October 10, 2011, from http://www.watergovernance.org/documents/WGF/Reports/Issue_sheets/WGF_Issue_Sheet_4_Water_and_Sanitation.pdf

UNICEF. (n.d). *Water, Sanitation and Hygien*. Retrieved October 15, 2011, from UNICEF: http://www.unicef.org/WaSH/index_statistics.html

UNU. (n.d). *Summary Analysis Sanitation and the Millennium Quantifying Water Supply,Development Goals*. Retrieved December 09, 2011, from http://www.inweh.unu.edu/Health/documents/WASHandtheMDGsresultssummarySept16_000.pdf

USAID. (n.d). *Ethiopia: Water and Sanitation Profile*. Retrieved October 16, 2011, from www.wsscc.org/sites/.../usaid_ethiopia_water_and_sanitation_profile.pdf

WB,ADB and WSP. (2009, October). *Building Partnerships for Sustainable Water and Sanitation Services in Africa*. Retrieved December 13, 2011, from http://www.afdb.org/fileadmin/uploads/afdb/Documents/Generic-Documents/Building%20Partnerships%20for%20Sustainable%20Water%20and%20Sanitation%20Services%20in%20Africa.pdf

WSP. (n.d). *Water Supply and Sanitation in Ethiopia*. Retrieved December 14, 2011, from www.wsp.org

WVE. (2012). "Impacts of Fessa Water Supply and Saniation Project in Village" Gache. 2.

WVE. (2006). *Redesign Document (2006-2010).* Agena.

WVE. (2011). *World Vision Ethiopia Guraghe Area Development Program Redesign Document.* Agena.

Annex –A: Survey questionnaire for potable water users

Dear respondent, my name is Israel Yigezu and I am a post graduate student at Hawassa University that I am being conducting a survey on potable water services in your area for educational purpose. I would very much appreciate your participation in this survey since the information you provide will help me much for the accomplishment of the study and for the better implementation of drinking water services in your *woreda.* The assessment question takes up to 15 minutes to complete. Whatever information you provide will be kept strictly confidential and will not be shown to other persons.

NB: Participation in this study is voluntary and you can choose not to answer any individual question or all of the questions. However, I hope that you will participate in this assessment since your views are very important for the completion of the study. Do you have any question about the assessment? May I begin the interview now?

1. Kebele: 1) Sheramo 2) Shebreden 3) Desene 4) Agena 01 5) Weyira Debane 6) Wasamar
2. Sex: 1) Male 2) Female
3. Age : 1) 18-25 2) 26-30 3) 31-35 4) 36-40 5) 40-50 6) 51 and above
4. Educational Status: 1) Illiterate 2) Grade 1- 5, 3) Grade 6-8, 4) 9-10 5) 11-12, 6) college graduate
5. Marital Status: 1) Single 2) Married 3) Divorced 4) Spouse died
6. Number of your family members: 1) 2-3 2) 4-5 3) 6-7 4) 8-10 5) 11-14 6) 16 and above
7. Occupation: 1) Farmer 2) Merchant 3) Artisan 4) gov't employee 5) student 6) Jobless
8. How do you rate drinking water services to you? 1) good 2) very good 3) Excellent 4) Bad 5) very bad
9. What type of water do you use for drinking? 1) River 2) pipe 3) pond 4) hand dug well 5) Rain 6) Pump 7) spring
10. Does your water source function throughout the year? 1) No 2) weak in winter 3) Sufficient 4) lost some months 5) Constant
11. Do you think drinking water usage in your area is efficient? 1) Yes 2) no

12. If your answer for question above is no, why? ____________________

13. Does the water you drink have any quality problem? 1) yes 3) no 3) don't know

14. If your answer for question above is yes, what type of water quality problem have you faced?

    1) Oder 2) Taste 3) Salty 4) Color 5) impurity

15. How much water in liter do you use per day for your household? ____________________

16. How many litters of potable water can you get per day? ____________________

17. How many minutes do you travel to collect water? ____________________

18. Is there problem in drinking water distribution mechanism? 1) Yes 2) no

19. If your answer to question above is yes, what are the problems? ____________________

20. Is there equity in drinking water supply? 1) If no, why? ____________________

21. Who are major drinking water actors in your area? 1) NGO, GO, FBO, Artisans, and CBO 2) GO and NGO 3) GO 4) Go, NGO and Community

22. Who has constructed the water you are using? 1) NGO 2) Government 3) Community 4) FBO 5) NGO, GO, FBO, Artisans, and CBO 6) none is constructed

23. How do you see the coordination between water actors in your area? 1) Very poor 2) Poor 3) sufficient 4) Very good 5) Excellent 6) Have no knowledge

24. Does water related decisions made by actors are accountable for the water users?
    1. Woreda Water, Minera and Energy Office? 1) Accountable 2) Not Accountable
    2. World Vision Ethiopia? 1) Accountable 2) Not Accountable
    3. Kalehiwot? 1) Accountable 2) Not Accountable
    4. WaSH Committee? 1) Accountable 2) Not Accountable

25. If your answer for question above is no, why do you think? ____________________

26. Do water actors are responsive to water related questions in your area?
    1. Woreda Water, Minera and Energy Office? 1) Responsive 2) Not responsive
    2. World Vision Ethiopia? 1) Responsive 2) Not responsive
    3. Kalehiwot? 1) Responsive 2) Not responsive
    4. WaSH Committee? 1) Responsive 2) Not responsive

27. If your answer for question above is no, why do you think? ___________________________

28. Do decision made regarding water issue is communicated to the community? 1) Yes 2) no

29. Do government sector offices, NGOs, water committee and other water actors work closely in solving water related problems? 1) Yes 2) no 3) have no information

30. If your answer for question above is no, why do you think? ___________________________

31. Is there any water scheme constructed in your area? 1) yes 2) no 3) I have no information

32. If your answer for question above is yes, what type of water scheme is constructed? 1) Water point development 2) Deep well 3) Shallow well 4) Pipe extension 5) traditional underground

33. Which body is responsible for the regulation of water scheme? 1) *woreda* water Bureau 2) Woreda health 3) Water committee 4) KA administration 5) Woreda administration 6) Other

34. Which body is responsible for the monitoring and evaluation of post water scheme construction? ___________________

35. Is there any identified planning procedure for Potable water projects? 1) Yes 2) no

36. If no, why do you think?____

37. What sort is water point you use in your area? 1) protected 2) unprotected

38. Is there any responsible body that penalizes deliberate water wastage and scheme damage in your area? 1) yes 2) no

39. If your answer above is yes, which body is responsible for? ___________________________

40. If the water you drink is pipe/developed water, do you pay water fee? 1) yes 2) No

41. If your answer for question above is no, are you willing to pay if needed? 1) yes 2) no

42. For which investment is it better to use water fee? 1) improve supply 2) improve management 3) Improve quality 4) Other_______________________________________

43. Have you taken any water related trainings and awareness? 1) Yes 2) no

44. If your answer for question above is yes, who gave that training? 1) Government 2) NGO 3) FBO 4) CBO 5) volunteer individuals

45. Have you ever informed about the decision made on water issues in your area? 1) Yes 2) no

46. If no, what do you think is the problem? ___________________________________

47. Do you participate in the construction of water scheme in your area? 1) Yes 2) No

48. If yours question above is yes, in what way have you participated? 1) Money 2) construction material support 3) labor supply 4) professional assistance

49. In what way the cost of water scheme maintenance financed? 1) through fee collected from user 2) from Goven't 3) from community 4) from NGOs 5) From private

50. What other challenges have you observed on drinking water services in your area? Please write it

____

____

51. What are the opportunities of drinking water supply in your area?

____

____

52. What do you recommend for the better safe water services in the woreda?

____

____

Annex-B: Interview checklist for potable water actors

Dear respondent, my name is Israel Yigezu and I am a post graduate student at Hawassa University that I am being conducting research paper on potable water services in your area for educational purpose. I would very much appreciate your participation in this survey since the information you provide will help me much for the accomplishment of the study and for the better implementation of drinking water services in your *woreda.* The assessment question takes up to 15 minutes to complete. Whatever information you provide will be kept strictly confidential and will not be shown to other persons.

NB: Participation in this study is voluntary and you can choose not to answer any individual question or all of the questions. However, I hope that you will participate in this assessment since your views are very important for the completion of the study. Do you any question about the assessment? May I begin the interview now?

1. What is the water supply capacity in liter/sec in the *woreda*?
2. Do you thing drinking water usage in Edja *woreda* is efficient? 1) Yes 2) no
3. If your answer for question above is no, why do you think?
4. Does the water you drink have any quality problem? If yes why?
5. What is the average minute that an individual travel to collect water in the *woreda*?
6. Is there problem in drinking water distribution mechanism? 1) Yes 2) no
7. If your answer to question above is yes, what are those problems
8. Is there equity in drinking water supply? 1) No, 2) Yes; if no, why?
9. Who are major drinking water actors in *woreda*?
10. What are the roles of the following drinking water actors in potable water management?
    1. NGOs____
    2. Water Bureau____

3. Water Committee____

4. Drinking water users____

11. Do you think that appropriate and timely decision is made on water service issues? 1) Yes 2) no

12. If your answer for question above is no, what is the reason?

13. Do water actors in the *woreda* are accountable to users? 1) yes 2) no

14. If your answer to question above is no, why?

15. How do you see the networking status between drinking water actors (NGOs, Woreda water Office, water committee, and the beneficiaries) in the *woreda*? 1) Not bad 2) good 3) very good 4) Excellent 5) weak 6) very weak.

16. If there is no good networking among actors, what do you think is the problem?

17. What needs to be done to improve networking and coordination between actors?

18. Does the drinking water users pay water fee in the *woreda*? 1) yes 2) No 3) pay at subsidized price

19. Is there any penalty mechanism in case of wastage and deliberate scheme break down? 1) yes 2) no

20. If your answer for question above is yes, which body is responsible for it?

21. Which body regulates drinking water schemes?

22. Do the water users participate on the process of water project? 1) yes 2) No

23. If your answer for question above is yes, at what stage the users participate?

24. What measure is used to increase water awareness among users?

25. Which part is responsible for maintenance report?

26. Who is responsible for the maintenance of water scheme in case of malfunctioning?
27. Do the beneficiaries or their representatives participate in water issue decision making process?
28. Which body is responsible for the monitoring and follows up of post water scheme construction?
29. Do the beneficiaries pay water fee?
30. If answer for question above is yes, for what purpose the fee used?
31. If the answer for question above is no, why?
32. I what way the cost of water scheme maintenance financed? 1) through fee collected from user 2) from Goven't 3) from community 4) from NGOs 5) From private
33. What are some of the other major challenges of drinking water services in the *woreda*?

____

34. What are some other opportunities of drinking water supply in the *woreda*?

____

35. What do you recommend for the better safe water services in the *woreda*?

____

____

Annex-C: The Focus Group Discussion Checklist

1. How much water a household use in liter/sec per day in your area?
2. Discuss the effectiveness of water usage in your village.
3. Discuss the quality of water you are using
4. Discuss the average minute that an individual travel to collect water in your area.
5. Discuss on the drinking water distribution mechanism in your area.
6. Discuss on the potable water distribution problems in your village
7. Discuss on the status of equity in water distribution.
8. Who are major drinking water actors in your area?
9. Discuss on the roles of the following drinking water actors on potable water management
    1. NGOs
    2. Water Bureau
    3. Water Committee
    4. Drinking water users
10. Do you think that appropriate and timely decision is made on water service issues?
11. If your answer for question above is no, what is the reason?
12. Discuss on the status of accountability water actors to users.
13. How do you see the networking status among drinking water actors in your area?
14. What needs to be done to improve networking and coordination between actors?
15. Does the drinking water users pay water fee?
16. Is there any penalty mechanism in case of wastage and deliberate scheme break down?

17. If your answer for question above is yes, which body is responsible for it?
18. Discuss on the body that regulates drinking water schemes in your village.
19. Do the water users participate on the process of water resource development project?
20. If your answer for question above is yes, at what stage the users participate?
21. What measure is used to increase water awareness among users?
22. Which part is responsible for maintenance report?
23. Who is responsible for the maintenance of water scheme in case of malfunctioning?
24. Do the beneficiaries or their representatives participate in water issue decision making process?
25. Which body is responsible for the monitoring and follows up of post water scheme construction?
26. Do the beneficiaries pay water fee? If yes for what purpose the fee used?
27. I what way the cost of water scheme maintenance financed?
28. What are some of the other major challenges of drinking water services in your area?

    ____

29. What are some other opportunities of drinking water supply in your village?

    ____

30. What do you recommend for the better safe water services in your village?

    ____

Annex-D Time schedule of the study

S/N	Phase	Major activities	Sub Activities	Required Time
1	Phase One	Review of related literatures	Any related studies conducted on Potable water management were searched and necessary ideas were incorporated in the study document to create background and context for the research problem. Documents related to the related Edja *Woreda* were also review to grasp enough knowledge about the area.	30 days
2	Phase Two	Field Study design	The study employs both qualitative and quantitative data collection method to generate information from respondents. Questionnaire was given for water and sanitation beneficiaries and government officials. Quantitative data was also collected from both government officials and other institutions. Guiding protocols for each of these key methods were designed.	10 days
3	Phase Three	Reconnaissance and rapport building visit to the study area	Selection of specific data collection in the study area, the identification of the respondents, group discussions, selection of field guides and assistant enumerators were conducted.	7 days
4	Phase Four	Data collection	Collection of field data through distribution and collection of questionnaires to the program beneficiaries and government officials a videotaping, photographing of some facts concerning the study topic was conducted.	30 days
5	Phase Five	Data Transcription	First, all the collected data will be ordered, organized, edited, cleaned, coded and reduction process will be made to make it ready for analysis. The coded data was entered in to the SPSS software.	15 days
6	Phase Six	Data feeding	The coded and edited data was entered in to the SPSS software.	10 days
7	Phase Seven	Preliminary Data analysis	Analysis of the raw data was undertaken using the statistical package for social science software	10 days
8	Phase Eight	Interpretation of the processed data and write up the report	Knowledge was generated from the processed data at stage 6 above and the generated knowledge was written in report form.	30 Days
Total Required Time for the project				142 days